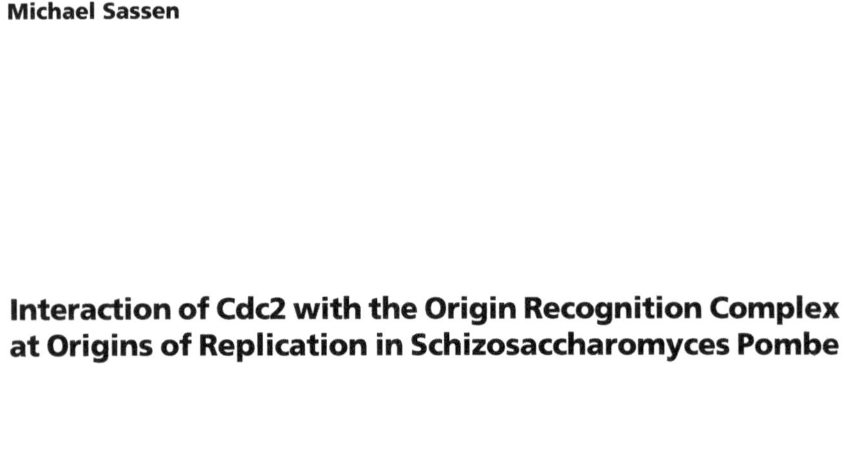

Michael Sassen

# Interaction of Cdc2 with the Origin Recognition Complex at Origins of Replication in Schizosaccharomyces Pombe

GRIN Verlag

**Bibliografische Information der Deutschen Nationalbibliothek:**

Die Deutsche Bibliothek verzeichnet diese Publikation in der Deutschen Nationalbibliografie; detaillierte bibliografische Daten sind im Internet über http://dnb.d-nb.de/ abrufbar.



**Imprint:**


Druck und Bindung: Books on Demand GmbH, Norderstedt Germany
ISBN: 978-3-638-65630-6

**This book at GRIN:**

http://www.grin.com/en/e-book/41889/interaction-of-cdc2-with-the-origin-recognition-complex-at-origins-of-replication

# Interaction of Cdc2 with the origin recognition complex at origins of replication in *Schizosaccharomyces pombe*

**Michael Sassen**

Technische Universität Bergakademie Freiberg

Fakultät für Chemie und Physik, 09596 Freiberg

Leatherwood at the Department of Molecular Genetics and Microbiology, State University of New York, Stony Brook, New York, 11794

# TABLE OF CONTENT

# Chapter 1

*"We are unlikely to ever know everything about every organism. Therefore, we should agree on some convenient organism(s) to study in great depth, so that we can use the experience of the past to build on in the future" (Huxley, 1869).*

# Introduction

## 1.1 *S. pombe* as a Model System

*Schizosaccharomyces pombe* functions as a suitable model system since it is easy and inexpensive to rear, has a convenient size, a short life cycle, and is genetically manipulable. As a unicellular eukaryote, the fission yeast *S. pombe* can exist either in a haploid or diploid state and possesses two different mating types ($h^+$ and $h^-$). The wildtype however is $h^{90}$, which means it can switch mating type. *S. pombe* shows a lot of similarity to *Saccharomyces cerevisiae*. However, the morphology of the two cells is different, with *S. pombe* being more rectangular than the circular-like *S. cerevisiae* cells (Hochstenbach, 1998).

*S. pombe* can undergo two different life cycles, either the vegetative (mitotic) cycle or the sporulation (meiotic) cycle, depending on the environment it is living in. These two cycles are shown in figure 2 with the change between the two occurring in cells at the $G_1$ stage of the mitotic cycle. Under laboratory conditions, given all nutrients required, *S. pombe* prefers the haploid state. This makes it a favorable organism for genetic research since it ensures that introduced mutations are not masked by another wildtype allele.

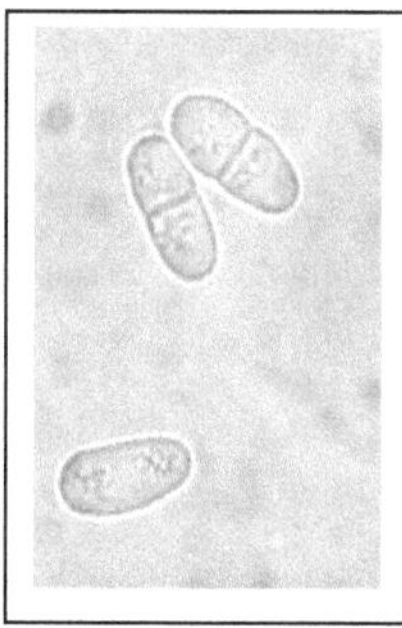

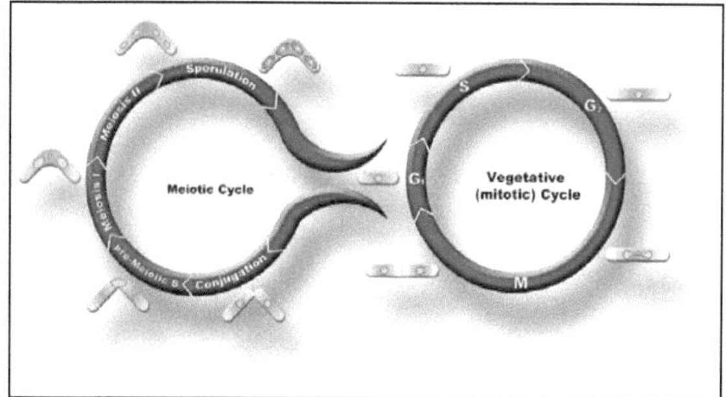


**Fig. 1: Left, picture of *S. pombe* cells.** At the top are two dividing cells in late Mitotic phase, showing the fission yeast typical septum at the point of cytoplasmic division. The lower cell is in early M phase, having its chromosomes already segregated.
**Fig. 2: Right, fission yeast cell cycle**. Diagrammatic representation of the *S. pombe* cell cycles with the interchange between the two occurring in G1 phase (Figure obtained and used with permission from Trevor Pemberton, University of Sussex).

## Eukaryotic DNA Replication - Once Per Cell Cycle Only

Transmission of genetic information from one cell generation to the next requires the accurate and complete duplication of each DNA strand exactly once before each cell division. Typically, this process begins with the binding of an "initiator" protein to a specific DNA sequence or "replicator". In response to the appropriate cellular signals, the "initiator" directs a local unwinding of the DNA double helix and recruits additional factors to initiate the process of DNA replication. This paradigm describes most of the currently tractable replication systems and, although derived from prokaryotic and viral systems, there is no compelling reason to doubt that it will apply to all eukaryotic organisms. In fact, the proteins that regulate replication are highly conserved from yeast to humans, including the origin recognition complex (ORC), which binds directly to replication origin sequences in fission yeast (Leatherwood, 1996) as well as in all other eukaryotes tested (Diffley, 2001;Kelly, 2000).

## The Origin Recognition Complex – a Closer Look

The origin recognition complex (ORC) plays a central role in initiation of DNA replication in eukaryotic cells. It interacts with origins of DNA replication in chromosomal DNA and recruits additional replication proteins to form functional initiation complexes. Competition binding experiments demonstrated that ORC binds preferentially to DNA molecules rich in AT-tracts, but does not otherwise exhibit a high degree of sequence specificity. As shown in figure 3, from its six subunits, labeled accordingly to their size Orp1 – Orp6, only Orp4 binds through its N-terminal nine repeats long AT-Hooks directly to its cognate origins. Interestingly, although the remaining five subunits of ORC were able to interact with Orp4 bound to DNA, they did not appear to have any sequence-specific DNA binding activity on their own, nor did they alter the interaction of prebound Orp4 with the origin DNA (Dutta, 1997).

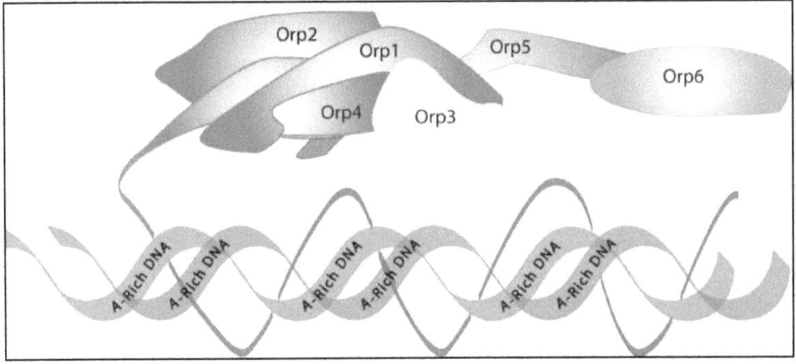


**Fig 3: The Origin Recognition Complex**
Binding of *S. pombe* ORC (SpORC) to its cognate origins involves a clearly distinct mechanism from the one in *S. cerevisiae* involving nine repeats of an AT-hook motif found uniquely at the N-terminus of the *S. pombe* Orp4 subunit. A discrete binding site has not been identified, however recent studies indicate that SpORC recognizes stretches of A-rich DNA.

In most eukaryotic systems, binding of ORC to DNA is intimately linked to ATP binding and hydrolysis by ORC (Dutta, 1997). But in *S. pombe,* however, it is not. Although ATP binding is not substantial for a stable DNA binding by *S. pombe* ORC it might be needed for the recruiting of other proteins necessary for the initiation of replication. The function as an ATPase of ORC remains to be determined for all eukaryotes. Possible events that could be coupled to ATP hydrolysis include the assembly of protein complexes at the origin, changes in the origin DNA or the associated protein complexes during the initiation of DNA replication, or disassembly of origin-protein complexes after initiation has occurredThe function of ORC in eukaryotic cells is to select genomic sites where pre-replicative complexes (pre-RCs) can be assembled. Subsequent activation of these pre-RCs results in bi-directional DNA replication that originates at or close to the ORC DNA binding sites. Since ORC binding to DNA is the first step in pre-RC assembly, ORC determines where DNA replication can begin by binding specific DNA sites in the genome.

Initiation sites for DNA replication are distributed throughout all eukaryotic genomes. This insures that genomes can be duplicated in a few minutes (e.g. rapidly cleaving embryos of flies, frogs, and fish) to a few hours (e.g. mammalian cells) without having to change the basic replication fork mechanism by which DNA is replicated. In yeast, flies and mammals, cis-acting sequences (replication origins) determine where DNA replication will begin (DePamphilis 1999, Altman and Fanning 2001, Lu 2001, Bell and Dutta 2002), but with the exception of the small (app. 0.1 Kb) replication origins in *S. cerevisiae*, no consensus sequence has been identified that is required for replication. Replication origins in fission yeast, flies and mammals are five to 20 times larger than those in budding yeast. They frequently contain large AT-rich regions and genetically identifiable sequences that are required for origin function. The activity of these sequences is often orientation or distance dependent.

**Interactions between ORC and DNA Replication Proteins – the Pre-RC**

ORC binding to the replicator is the first step in the establishment of a multifactor assembly called the pre-replicative complex (pre-RC, for review see Bell, 2002). But how does ORC find its way to the tiny motifs within the genome, given an average spacing between strong origin sequences of app. 20-30 Kb? It seems unlikely that the sequence specificity of ORC is sufficient for localizing these origin sequences all by itself. Studies of the interaction of ORC with origin DNA in the presence of other pre-RC components suggest that these factors may contribute to the specificity of ORC localization. The recruitment of these factors occurs in an ordered process: Cdc18 (*S. pombe* name; also known as Cdc6 in *S. cerevisiae*) and Cdt1 load first followed by Mcm2-7, which may function as the replicative helicase (Tye, 1999). As cells pass through the $G_1$ to S phase transition additional replication factors are recruited to the origin. These factors include Mcm10 and Cdc45, the three eukaryotic DNA polymerases, and the eukaryotic ssDNA binding protein, RPA (for review, see Bell, 2002). As with the components of the pre-RC, either cells or extracts lacking ORC fail to assemble these additional factors at the origin. Although it is clear that ORC is required to initiate the assembly of all of these factors at the origin, ORC has only been shown to interact directly with a small subset of these factors. The preinitiation complex is apparently fully assembled in $G_1$ phase, and yet initiation does not occur. Initiation is not triggered until two protein kinases become active. These kinases are Hsk1 (*S. pombe* name; also known as Cdc7 in *S. cerevisiae*) and Cdc2 (Vas, 2001).

### Cdc2 Control of the Cell Cycle in *S. pombe*

Cdc2 is a cyclin-dependent kinase (CDK), which is activated at different stages throughout the cell cycle mainly by the cyclins Cig2 and Cdc13. In *S. pombe* only one CDK, namely Cdc2 regulates the cell cycle (Moreno, 1989). Starting with no activity during $G_1$ phase it allows the pre-RC to be assembled in preparation for DNA replication. The progression into S phase is allowed by the raising activity of Cdc2. This is accomplished in part because Rum1, a Cdc2 inhibitor, is degraded and the cyclins Cic1 and Cig2, lacking in $G_1$, are synthesized and bound to Cdc2. At the same time the increased Cdc2 activity does not allow new formation of pre-RC and therefore prevents re-replication (Correa-Bordes, 1995). More cyclins accumulate during S phase and $G_2$ but their effect on Cdc2 is held in check by inhibitory phosphorylation by the Wee1 and Mik1 kinases. In order to trigger mitoses, the activity of Cdc2 becomes higher at the end of $G_2$ by its dephosphorylation via Cdc25 phosphatase and its binding to the mitotic cyclin Cdc13. At the very end of M phase Cdc2 is inactivated by expression of Rum1, destruction of Cdc13 and several other mechanisms (Lee, 1999; Booher, 1989).

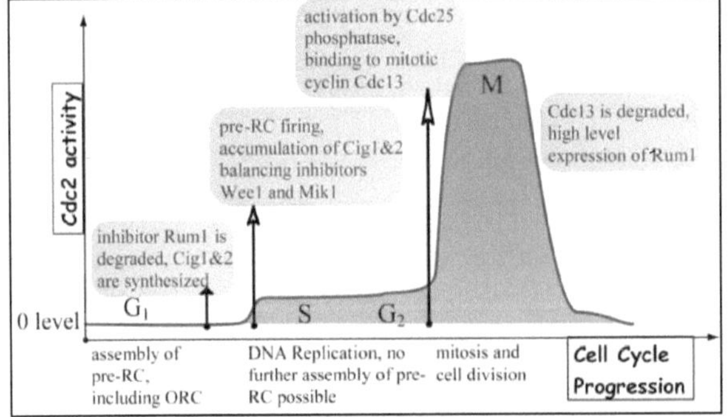


**Fig 4: Cdc2 activity throughout the cell cycle.** Cdc2 as the major cyclin-dependent kinase in *S. pombe* controls most of the cell's cycle related activity. Cdc2 in return is activated at certain checkpoints by different cyclins. The most important one is Cdc13 because, although it is not present at all stages throughout the cell cycle, it can substitute the other

cyclins and therefore activate Cdc2 all by itself. Cdc25 also has an activator function as it removes the inhibitory Tyr15 phosphorylation at Cdc2. Inhibitors including Wee1, Mik1, and Rum1 play the counterpart of activating cyclins and Cdc25. The whole cell cycle is a very precisely balanced system of inhibition and activation of Cdc2.

The function of Cdc2 is not limited to the control of DNA replication. It also takes part in control mechanisms of growth polarity, spindle pole body duplication, chromosome condensation, mitotic spindle functions, mitosis, and cytokinesis (Nurse, 1990).

**Cdc2 Control of Replication**

As shown in figure 5, the main CDK concentration in *S. pombe*, Cdc2, remains at a constant level throughout the cell cycle. Its regulatory function is established as its activity changes (Hendrickson, 1996). This change is triggered by Cdc2 binding to its cyclin partners Cdc13 and Cig2. Both cyclins act at different stages of the fission yeast cell cycle. Cdc13 formation peaks at the onset of M-Phase, causing a rapid rise of Cdc2 activity once it binds to it. Cig2 is responsible for raising the Cdc2 activity level at the beginning of S-Phase, allowing DNA replication to start. But whereas Cdc13 is essential, Cig2 is not. The whole cell cycle can be driven by Cdc13 alone (Nguyen, 2001). Therefore it is the cyclic destruction, reforming and binding to Cdc2 of the cyclins that state the internal clock of the cell cycle.

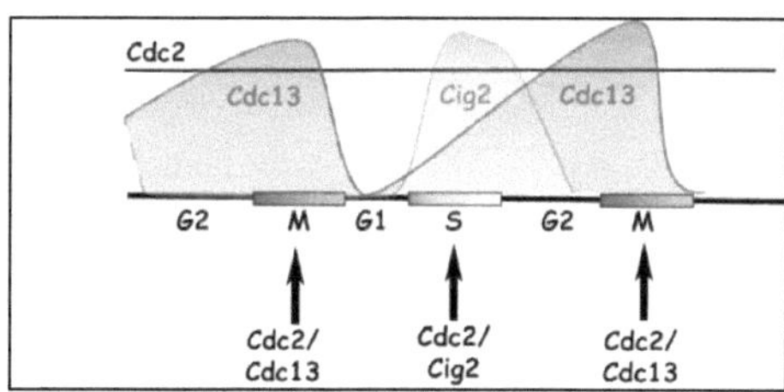


**Fig 5: Cyclic activation of Cdc2.** Cdc13 and Cig2, the main cell cycle driving cyclins, act at different stages of the fission yeast cell cycle. Both are B-type cyclins and activate Cdc2 by binding either during S- or M-phase. It is known that Cdc2 binds also to Cig1 and Puc1 cyclins, which have an overlapping role with Cig2 in the initiation of S phase and are therefore included under Cig2 activity.

Initiation of replication depends on Cdc2 kinase activity, and a number of origin binding proteins including Cdc18, Mcm4, Orc2, and Orc6 are targets for Cdc2-dependent phosphorylation in vitro and in vivo (Vas, 2001; Lopez-Girona, 1998). How can changes in Cdc2 activity control DNA replication? Proteins belonging to the MCM family have been identified as being associated with G1 chromatin, which is in a permissive state for replication. Those MCM family proteins are absent from G2 chromatin, which cannot normally replicate. Interestingly, a feature of these proteins is that they are hypophosphorylated in G1 and become phosphorylated during S phase. This phosphorylation is performed by Cdc2. The changes in phosphorylation are correlated with changes in MCM affinity for chromatin providing a strong link between the Cdc2 activity cycle and DNA replication licensing. Mcm4 on the other hand requires Orc1 and Cdc18 to function in a proper way, whereas Cdc18 itself is dependent on ORC (Kearsey, 2000; Lopez-Girona, 1998). These findings suggest ORC working as a "landing pad" to assemble the various replication proteins of the pre-RC at the origin of replication (Prasanth, 2002).

Although ORC binds constitutively to chromatin throughout the fission yeast cell cycle, the prereplicative complex only forms on origins during G1. In recent publications (Wuarin 2002 and citations therein) it is reasoned that a factor may stably bind ORC during G2 and early M phase to prevent reinitiation of S phase. Since the Cdc13 protein resides in the chromatin domain of fission yeast nuclei and is required to prevent re-replication, this factor might be the Cdc13/Cdc2 complex itself. In cells lacking Cdc13, association of Cdc2 to origins was abolished and conversely, the association of Cdc13 to origins was greatly diminished in *cdc2* ts-mutants. This mechanism is another example of the regulatory role of the Cdc13/Cdc2 complex. As with the phosphorylation of the Mcm proteins, prevention of re-replication by Cdc2 binding to ORC is mediated via association with the main cyclin Cdc13.

A third mechanism of DNA replication control by Cdc2 is the phosphorylation of the M phase checkpoint control factor Cdc18 (Ayté et al., 2001). Cdc2 may regulate replication licensing by catalyzing the destruction of nonchromatin bound Cdc18 protein. *cdc18* and *cig2* mRNAs and their corresponding proteins are both periodically expressed in synchronous endoredublicating cells lacking Cdc13 (*cdc13::ura4 nmt41-cdc13*). Cdc18 protein is degraded coincidently with the peak of Cig2 protein expression during each endoredublication cycle (Wuarin 2002). This suggests that Cig2/Cdc2 kinase prevents relicensing of origins both by catalyzing the periodic destruction of Cdc18 and by regulating its own synthesis. Once more this regulatory pathway shows the dependence of Cdc2 on its cyclin partner, here Cig2.

As previously found in vitro (Vas et al. 2001), Cdc2 phosphorylation of Orp2 appears to be one of multiple mechanisms by which Cdc2 prevents DNA rereplication in a single cell cycle. The cell cycle coupled regulation of replication is crucial for genome stability in every organism. I am particularly interested in the regulations that ensure genome stability because studying these pathways in yeast gives us insights into the mechanisms that are critical in mammalian cells to prevent cancer.

**Project Goal**

The goal of my project is to show an in vivo binding of Cdc2 to the origin recognition complex in *S. pombe*. The existing model of initiation of DNA replication is based on in vitro findings of a Cdc2 – ORC interaction in fission yeast (Leatherwood 1996). To show this interaction in vivo, a working chromatin immunoprecipitation (ChIP) assay had to be further developed to be employed routinely in the lab. The ChIP assay is a valuable method to show DNA – protein as well as protein – protein interactions in vivo. Existing protocols had to be modified and extended as well as new parts were to be added. Especially measures to enhance the stringency of the assay, such as the CsCl gradient centrifugation had to be implemented. Additionally more sensitive detection methods, such as optimized PCR programs and real time PCR detection should be employed.

To ensure the reliability of the ChIP assay, previously published ChIP experiments (reviewed in Gavin KA et al. 1995) were redone and the results were compared as control. It was to show that proteins of the origin recognition complex bind to origins of replication in *S. pombe*.

# Chapter 2

# Materials

## 2.1 Sources of Used Chemicals, Enzymes, and Kits.

American National Can, USA

BioRad Laboratories, Hercules, CA, USA

Fisher Scientific, Pittsburgh, PA, USA

J.T. Baker, Phillipsburg, NJ, USA

Mettler Toledo, Switzerland

Molecular Probe, Eugene, OR, USA

Nortech Laboratories Inc., Farmingdale, NY, USA

Pharmacia Biotech, Uppsala, Sweden

Q-Bio gene, USA

Qiagen Inc., Valencia, CA, USA

Roche Diagnostics GmbH, Mannheim, Germany

Sigma Chemical Co., St. Louis, MO, USA

VWR Scientific, USA

## 2.2 *S. pombe* Strains

| Strain | Genotype of interest | Complete Genotype |
|---|---|---|
| F14 | Wildtype | h+ ade6-M210 his7-366 leu1-32 ura4-D18 |
| JLP285 | *orp2-GFP* | h? GFP-orp2, ade+(8) his+ leu+ ura+ |
| JLP464 | *cdc2-GFP* | h+ *cdc2::cdc2-GFP* ade6-M210 his+ leu1-32 ura4-D18 |
| JLP479 | *cdc2-GFP, cdc18* ts | h? *cdc2::cdc2-GFP cdc18* ts ade6-M216 his+ leu+ ura4-D18 |
| JLP910 | *cdc2-GFP, orp2-2* ts | h? *cdc2::cdc2-GFP orp2-2* leu1-32 ura4-D18 |
| JLP912 | *cdc2-GFP, orp2-7* ts | h? cdc2::cdc2-GFP *orp2-7* ade6-M216 leu1-32 ura4-D18 |
| JLP955 | *cdc2-4064-GFP* ts | h- *cdc2::cdc2-4064-GFP* ade6-M210 his7-366 leu1-32 ura4-D18 |

Table 1: *S. pombe* strains

## 2.3 Oligonucleotides

| Oligonucleotide | Sequence | Description |
|---|---|---|
| 9F | ACAAAACATGTGAGGAAAGCATG | origin-specific primer |
| 9R | CTCTCCACCTACTGTATTCCG | origin-specific primer |
| 17F | CGTATTCAATTGTCAGAGGTGAA | non origin-specific primer |
| 17R | ACATCCTTGGCAAATGCTTTCG | non origin-specific primer |

Table 2: Oligonucleotides

The important 3' region for specific priming is free of secondary structures, repetitive sequences, palindromes, and highly degenerated sequences.

## 2.4 Solutions and Yeast Media

All used chemicals were of analytical quality. The content of kit solutions can be obtained from the kit's user manual.

2.5 M Glycine
37 % Formaldehyde
70 % Ethanol
Anti-GFP antibody (Molecular Probe)
Bradford reagent (BioRad Laboratories)
Glass beads (Sigma 212-300 microns)
Protein A coupled to agarose beads (Pharmacia Biotech)
Qiaquick PCR Purification Kit #28104
LaRoche FastStart High Fidelity PCR System Kit #03553426001

**Yes Medium** 5 g Yeast Extract, 30 g Glucose, 0.19 g Adenine, 0.19 g Histadine, 0.19 g Leucine, 0.19 g Uracil. Distilled water ad 1000 ml, for plates add 20 g agar/L

**TE** 10mM Tris HCl (pH 8.0), 1mM EDTA (pH 8.0)

**PBS** 8 g NaCl; 0.2 g KCl; 1.15 g $Na_2HPO_4$* $7H_2O$; 0.2 g $KH_2PO_4$, $H_20$ ad 1 l

**ChIP Lysis Buffer (CLB)** 50 mM HEPES (pH 7.5); 140 mM NaCl; 0.1 % Sodium Deoxycholate; protease inhibitors are added fresh to a final concentration of 10x of 100x Benzamidine, 500x Pefabloc, and 1000x Pepstatin / Leupeptin

**ChIP Lysis Buffer + Triton (CLBT)** 50 mM HEPES (pH 7.5); 140 mM NaCl; (1 % Triton X100); 0.1 % Sodium Deoxycholate; Protease inhibitors are added fresh to a final concentration of 10x of 100x Benzamidine, 500x Pefabloc, and 1000x Pepstatin / Leupeptin

**PMSF** 0.2 M in Isopropanol

**Solution A** 4 % SDS, 20 mM Tris (pH 7.6), 150 mM NaCl, 2 mM EDTA (pH 8.0); 10x protease inhibitors

**Solution B** 2.2 % Triton X-100, 20 mM Tris (pH 7.6), 150 mM NaCl, 2 mM EDTA (pH 8.0), 10x protease inhibitors

**ChIP Lysis Buffer (High Salt)** 50 mM HEPES (pH 7.5); 500 mM NaCl; 1 % Triton X100; 0.1 % Sodium Deoxycholate

**ChIP LiCl Wash Buffer** 10 mM Tris (pH 8.0); 250 mM LiCl; 0.5 % NP-40; 0.5 % Sodium Deoxycholate; 1 mM EDTA

**ChIP Elution Buffer** 50 mM Tris (pH 8.0); 1 % SDS; 10 mM EDTA

**TEE-Glycerol Buffer** 5 % glycerol; 1 mM EDTA (pH 8.0); 0.5 mM EGTA (pH 7.5); 10 mM Tris-HCL (pH 8.0)

**CsCl-Gradient Solution** 10 mM Tris-HCl (pH 8.0); 1 mM EDTA; 0.5 mM EGTA; 0.5 % Sarkosyl (Sodium Lauryl Sarcosinate); 567.8 mg CsCl/ml

## 2.5 Equipment

Spectrophotometer (Hitachi U-1100)
Centrifuge (DuPont Sorvall RC-5B)
Bead beater (Biospec)
Sonicator (Ultrasonic W-220 F)
Vortex (Fisher Vortex Genie 2)
Waterbath (Fisher Isotemp 205)
PCR machine: MiniCycler (MJ Research)

# Chapter 3

# Methods

## 3.1 Growth of *S. pombe* strains

Strains F14, JLP285, and JLP464 were cultivated from freezer cultures on YES plates at 36 °C for 36 to 48 hours. A liquid 100 ml pre-culture of YES was inoculated with a single colony and shaken in an air shaker at 36 °C, 220 rpm. The main batch was inoculated with pre-culture cells, aiming for 1.0 OD after 16 hours growth. Temperature sensitive strains JLP479, JLP910, JLP912, and JLP955 were grown at their permissive temperature (28 °C) in the same way. Shifting to their restrictive temperature (36.5 °C) for 4 hours was performed rapidly, using a pre-heated waterbath. To ensure the arrest of the cells the septation index was determined.

*S. pombe* cell cultures for experiments can be stored at 4 °C for 2 weeks on YES plates. Viable cells for new plates can be obtained from YES plates up to 5 weeks.

## 3.2 Chromatin Immunoprecipitation (ChIP)

Protein-DNA complexes formed on chromosomes are key platforms for the initiation of replication in every cell. A variety of trans-acting proteins directly or indirectly interact with cis-acting chromosomal DNA sequences and form functional complexes that control many nuclear events, such as transcription, chromosome segregation, and - as in this projectwork described – replication. The ChIP method provides an advantageous tool for investigating these protein-DNA complexes. Whereas other methods, such as the Southern-Western analysis and gel shift assays, analyze direct interactions between protein and DNA in vitro, the ChIP method is suitable for detecting the interactions between proteins of interest and DNAs with known sequences in vivo. The first step of ChIP is fixation of living cells by formaldehyde. As shown in figure 6, rapid fixation takes place between proteins and DNA, leading to the conservation of near-native chromatin structures. Extracts prepared from fixed cells are sonicated to reduce the size of DNA fragments and used for immunoprecipitation.

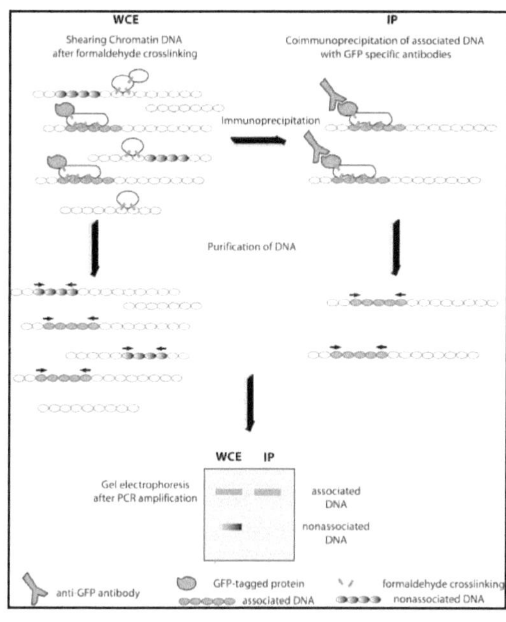


**Fig. 6: ChIP principle.**
The principle of the ChIP assay used to identify protein-DNA association *in vivo*. Living fission yeast cells are treated with formaldehyde to fix chromatin structures, then whole cell extracts (WCE) are prepared and sonicated to shear and solubilize the crosslinked protein-DNA and protein-protein complexes. A selective precipitation of the protein of interest fused to Green Fluorescence Protein (GFP, green) by an anti-GFP antibody (green Y) results in the co-immunoprecipitation of associated DNA (IP). Cross linking is reversed and the DNA is purified from both the WCE and IP samples. The precipitated DNAs are subjected to PCR amplification followed by gel electrophoresis to identify the sequences that were associated to the GFP tagged protein (orange). Nonassociated DNA sequences are shown in grey.

Reaction I

$$H_2N-CH(COOH)-(CH_2)_4-NH_2 + H_2C=O \rightleftharpoons H_2N-CH(COOH)-(CH_2)_4-NH-CH_2OH$$

Reaction II

$$H_2N-CH(COOH)-(CH_2)_4-NH-CH_2OH + H_2N-CH(COOH)-(CH_2)_4-NH_2 \underset{H_2O}{\rightleftharpoons} H_2N-CH(COOH)-(CH_2)_4-NH-CH_2-NH-(CH_2)_4-CH(COOH)-NH_2$$

**Fig. 7: Chemical crosslinking of DNA and proteins by formaldehyde.**
Formaldehyde (HCHO) is a very reactive dipolar compound in which the carbon atom is the nucleophilic centre. Amino and imino groups of proteins (e.g., the side chains of lysine) and of nucleic acids (e.g., cytosine) react with formaldehyde, leading to the formation of a Schiff base (Reaction I). This intermediate can react with a second amino group (Reaction II) and condenses. Crosslinks may be reversed by heating in Tris-HCl-containing buffers. This leads to a drop in pH and protonation of amino groups, thus forcing the equilibrium in the reverse direction. Shown is the crosslinking between the side chains of two lysines.

Co-immunoprecipitated, fragmented chromosomal DNAs are purified and the target sequence is amplified by polymerase chain reaction (PCR) and can be analyzed via gel electrophoresis. Note that ChIP can detect the indirect association of proteins with DNA, as the fixation procedure stabilizes large protein complexes that bind DNA. Once the PCR conditions are determined, ChIP permits the analysis of multiple DNA sequences by simply preparing PCR primers for the sequences of interest.

Cells of the desired strains were grown to an $OD_{600}$ of approximately 1.0 and the proteins were crosslinked to the DNA using 37% formaldehyde for 1 hour. After addition of glycine to stop the fixation, cells were washed with PBS and harvested. The protease inhibitors Benzamidine, Pefabloc, Pepstatin, and Leupeptin were added to prevent the protein complex from being digested. Lysates were prepared by adding glass beads (212-300 microns) to the cells and using a bead beater. To shear the genome into pieces of app. 1.3 Kbp length the samples were sonicated in the presence of the detergent Triton-X.

For the immunoprecipitation (IP) the protein concentration was determined, using the Bradford assay as advised (Bio-Rad Laboratories). Protein A (protA) bound to agarose beads and anti-GFP antibody were prebound and washed with ChIP Lysis Buffer (CLB). After combining the antibody complex with the samples several washing steps followed, using CLBT, TE, and PMSF. Whole cell extracts (WCE) were aliquoted as input controls. IPs were eluted by exposing them to increasing detergent conditions as well as heat, leading to the removal of protA/anti-GFP/GFP-binding. The protA-GFP and the protein-DNA complex were immunoprecipitated again, which increases the specificity of the assay. IPs were washed with buffers containing increasing salt concentrations (high salt CLB, LiCl CLB, TE). To remove the DNA-protein crosslinking, samples were placed at 65 °C for 18 hours.

DNA was extracted using the Qiagen DNA purification kit, amplified via PCR, resolved by gel electrophoresis, and visualized by ethidium bromide staining. For some experiments real time PCR (LaRoche FastStart High Fidelity PCR System Kit) in combination with melting curve analysis was conducted taking advantage of the LaRoche Light Cycler.

This protocol has some minor modification to the protocol described in (Strahl-Bolsinger, 1997) and was obtained from K. Flick (The Scripps Research Institute).

### 3.3 Application of ChIP to *S. pombe* mutant strains

To ensure that the selected fission yeast mutants show the desired properties, several control mechanisms have been implemented at various stages of the experiments. Since the proteins investigated in this study are essential, temperature sensitive (TS) strains with mutations in the gene of interest have been constructed. For most of the TS mutant strains (see table 1) it is crucial to ensure a complete arrest of the cell cycle. Once a culture is shifted to its restrictive temperature, in this case 36.5°C, no further cell cycle progression is possible because one or more essential proteins lose their native conformation and therefore their function.

As shown in figure 8, growing cells can be detected under the light microscope by formation of the fission yeast typical septum before cytoplasmic division occurs. The septation index indicates the percentage of cells that form a septum. A culture is considered arrested when the septation index is below 1%.

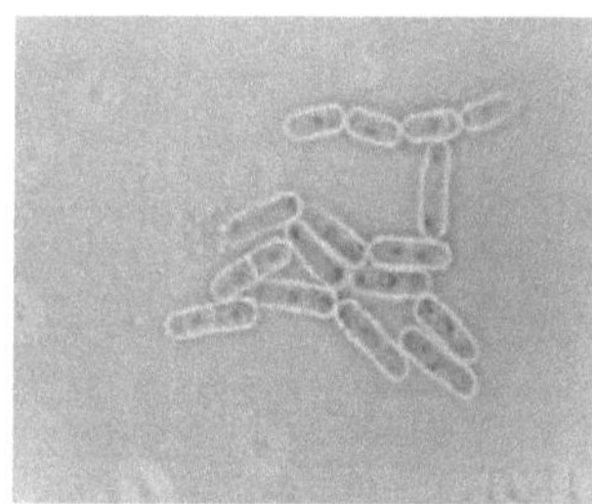

**Fig. 8: Arrested cells at restrictive temperature.**

Cells were taken after a four hour shift to their restrictive temperature. The magnification is 1000x, using a Zeiss Axioskop microscope and a Nikon Coolpix955 digital camera. The cells shown are from the temperature sensitive *cdc2* mutant JLP955 (cdc2-4064TS), where it is known that it does not arrest completely.

The next crucial step is the breaking of the cells to obtain the DNA. To ensure getting a yield of DNA as high as possible, cells must be broken thoroughly. Optimum results can be achieved having more than 90% of the cells broken. Figure 9 shows broken fission yeast cells of the temperature sensitive strain JLP910 after treatment with the bead beater. Similar results of breakage were obtained for all strains investigated. This ensures the highest possible yield of protein and DNA for the next experimental steps.

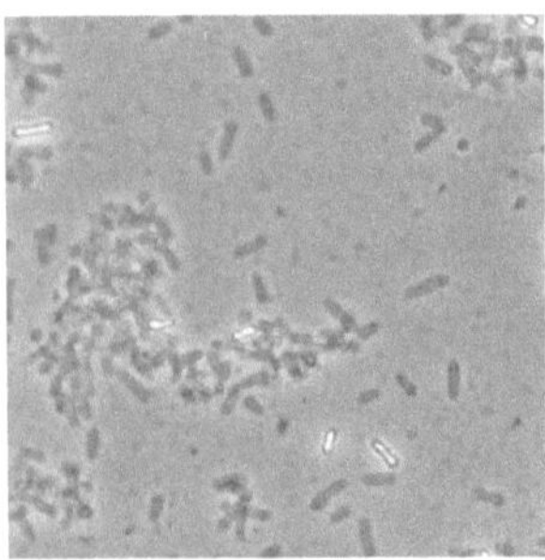

**Fig. 9: Broken cells JLP910.**

After 3 x 1 minute treatment in a bead beater the vast majority of cells is broken. The cytoplasm leaks and the remaining residues of cell debris cause the light to be reflected differently than in cells containing cytoplasm. Under the light microscope broken cells look black, whereas intact cells are bright.

To shear the genome into smaller pieces, the samples were sonicated in 16 bursts of 15 seconds each. As shown in figure 10, genome fraction-size is a function of sonication time. Although longer sonication bursts lead to smaller chromatin pieces, excessive sonication also heats up the cell lysate rapidly and leads to protein denaturation and degradation. To get a reasonable resolution and maximum yield of DNA-protein complexes at the same time, the optimum sonication time was determined to 16 x 15 seconds. Unsheared genome size reaches from 3 Kb to 8 Kb fragments, equally distributed (data not shown). After sonication most of the genome is broken into fragments of 1.3 Kb. DNA at this size can be reliably pulled down via chromatin IP.

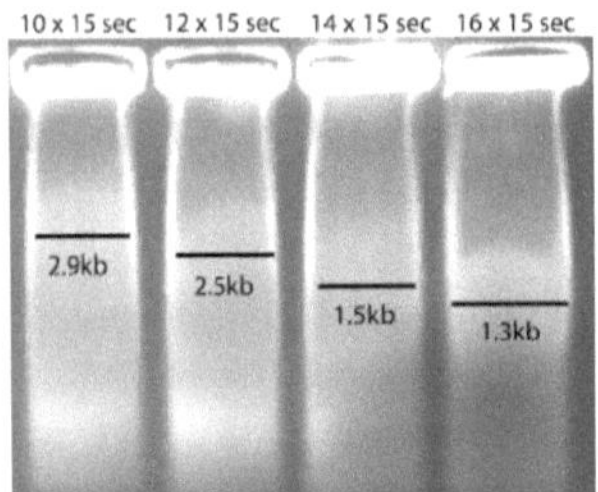


**Fig. 10: *S. pombe* DNA after sonication treatment.**
Cells were broken and sonicated, using the Ultrasonic W-220 F Sonicator with a micro-tip. After pelleting cell debris, DNA was purified and extracted via phenol/chloroform assay and ethanol precipitation as described previously (Sambrook, 1989). Fragments were resolved in TE and analyzed via agarose gel electrophoresis.

The origin-specific primers amplify a region of the ARS3001 sequence in *S. pombe* which is known to function as origin of replication. The non-origin primers were chosen to amplify a region of the genome that is not known for any replication related activity. To obtain a good resolution in a pull down assay like chromatin immunoprecipitation, two requirements have to comply: First, the sequence desired to pull down must be sufficiently far away from other *cis*-acting sequences. Second, the size of the sheared DNA must be small enough. Compared to the average size of sheared DNA after sonication of 1.3 Kb, the arrangement of origin and control sequences which are 2.7 Kb and 8 Kb apart, allows a reasonable resolution in the ChIP assay.

To obtain the protein concentration in each sample, the highly specific and very sensitive Bradford assay was used. The principle of this assay relies on an absorbance shift in Coomassie Brilliant Blue G-250 (CBBG) when bound to arginine and aromatic residues. The anionic (bound form) has absorbance maximum at 595 nm whereas the cationic form (unbound form (Sambrook, 1989)) has an absorbance maximum at 470 nm. The assay is monitored at 595 nm in a spectrophotometer, and thus measures the CBBG complex with the protein. The protein concentration shows slight variations from sample to sample and from experiment to experiment. To be able to compare levels of protein-DNA interaction, it was worked constantly with 5 mg of protein per IP.

### 3.4 CsCl-Gradient Centrifugation

The CsCl gradient is a type of density gradient centrifugation, a lab technique commonly used to separate or purify nucleic acids. The cesium chloride forms a density gradient (highly dense at the bottom, thinnest at the top), and the different nucleic acids separate along the gradient according to their buoyancies in different densities. This technique was used to eliminate nonchromatin-bound crosslinked proteins, together with naked DNA and RNA.

Where indicated the cesium chloride gradient was implemented after preparation of cell lysate for ChIP experiments. To purify the prepared cell lysate, it was dialyzed three times against 1 l of TE at 4 °C. Samples were adjusted to 0.5 % Sarkosyl and gently swirled at room temperature. Cell debris was pelleted by centrifugation at 16,100 *g*. The clear supernatant was mixed with CsCl, and using the refractive index, the density was set to 1.42 g/cm$^3$. Nonchromatin-bound crosslinked proteins, together with naked DNA and RNA, are eliminated by cesium chloride isopycnic centrifugation at 151,300 *g* for 72 hours. The formed gradient was fractionated from the bottom. To identify peak DNA-chromatin-containing fractions, the absorption at 260 nm, 280 nm, and the refractive index were taken. Fractions of the desired DNA-Protein crosslinked complexes were pooled and dialyzed against TEE-glycerol buffer at 4 °C. The purified DNA-Protein complexes were subjected to chromatin immunoprecipitation as described.

The decision, which fractions of the CsCl gradient to pool is based on three criteria. First, the yield of crosslinked DNA should be as high as possible. Although there will be always some loss, this criteria asks for the maximum of possible pooling. Second, the peak DNA-chromatin-containing fractions correspond to a density of approximately 1.39 g/cm$^3$. Pooling fractions with high differences to this density makes it more likely to contaminate the sample with free DNA or uncrosslinked protein. Third, the protein / DNA ratio (absorbance ratio 260 nm to 280 nm) for crosslinked protein / DNA complexes is between 1.8 and 1.95. A ratio above 1.95 indicates free DNA and RNA, and a ratio below 1.8 indicates free protein (Orlando 1997). For the later analysis free DNA states a serious problem and may falsify the results, whereas free protein does not pose a danger. Therefore one values criteria number one higher when it comes to decide what fractions to include from the top of the gradient right side in figure 13). To exclude the possibility of pooling fractions with free DNA strict emphasis is laid on criteria 2 and 3.

## 3.5 Phenol / Chloroform Extraction and Ethanol Precipitation

The phenol/chloroform extraction is a routinely used method for the removal and denaturing of proteins in nucleic acid solutions. Proteins are soluble in phenol/chloroform and can be separated from the water-soluble DNA by the formation of two layers.

300 µl of DNA extract were mixed with equal amounts of phenol/chloroform and centrifuged at 16,100 *g*. The upper phase containing DNA was transferred and after increasing the salt concentration, 2 volumes of 95 % Ethanol (-20 °C) were added and incubated at -80 °C. DNA was pelleted and washed with 70 % Ethanol (-20 °C) (Sambrook et al.). The remaining pellet was air dried, resuspended in TE, and analyzed via gel electrophoresis.

## 3.6 PCR and Real-time PCR

The PCR (polymerase chain reaction) technique is a primer extension reaction for amplifying specific nucleic acid sequences in vitro. The use of the thermostable *Taq*-polymerase allows the dissociation of newly formed complimentary DNA and subsequent annealing of primers to the target sequence with minimal loss of enzymatic activity. PCR will allow a short stretch of DNA to be amplified to about a million fold so that one can determine its size, nucleotide sequence, etc. The particular stretch of DNA to be amplified, called the target sequence, is identified by a specific pair of DNA primers, oligonucleotides usually about 20 nucleotides in length.

**PCR reaction mix:**

| | |
|---|---|
| DNA | 75 ng (genomic DNA) |
| | 100 ng (ChIP DNA) |
| Denville Buffer (10 x) | 1 x |
| BSA (Bovine Serum Albumin 1 mg/ml) | 0.3 µg |
| dNTP mix (2 mM) | 1.2 pM |
| *Taq*-polymerase (5 U/µl) | 0.3 U |
| Primer mix (10 pmol/µl) | 15 pmol each |
| $MgCl_2$ (25 mM) | 4 mM |
| $dH_2O$ | ad 30 µl |

**PCR program:**

4 min 92 °C
1 min 92 °C }
1 min 56 °C } 25 x
2 min 72 °C }
7 min 72 °C

To perform the real-time PCR and melting curve analysis, the LightCycler instrument from LaRoche was used together with the LaRoche PCR preparation kit. The principle of real-time PCR is the combination of the thermo-cycler component and a fluorimeter component for the in-process detection of amplified DNA products. A sequence-independent fluorescent dye (SYBR Green I) is added to the PCR template and its inherent fluorescence is enhanced when it binds to the minor groove of dsDNA products as they are synthesized. Thus the increase in SYBR Green I fluorescence, when measured at the end of each elongation cycle, indicates the amount of PCR product formed during that cycle. To amplify the origin of replication sequences, the origin specific primer pair 9F/R was used. For the non origin control sequence, the primers 17F and 17R were used (see table 2).

**Real-time PCR reaction mix:**

| | |
|---|---|
| DNA | 50 ng (genomic DNA) |
| | 75 ng (ChIP DNA) |
| Forward Primer (10 µM) | 0.5 µM |
| Reverse Primer (10 µM) | 0.5 µM |
| $MgCl_2$ (25 mM) | 4 mM |
| LightCycler kit | 1 x |
| $H_2O$ | ad 20 µl |

**Real-time PCR program:**

10 min 95 °C

10 sec 95 °C
05 sec 56 °C } 45 x
10 sec 72 °C

Melting curve acquisition

30 sec 40 °C

Working with real-time PCR as a tool to amplify DNA fragments reproducible, requires several calibration experiments. As shown for the origin specific 9F/R primer pair in figure 11, among the various concentrations of $MgCl_2$, a total concentration of 3mM yielded the best signal to background relation. In an ongoing genomic DNA titration it was found that even at a 1:125,000 dilution of genomic DNA the signal was still clearly above background (figure 12). More parameters were tested this way (results not shown) and led to optimized settings.

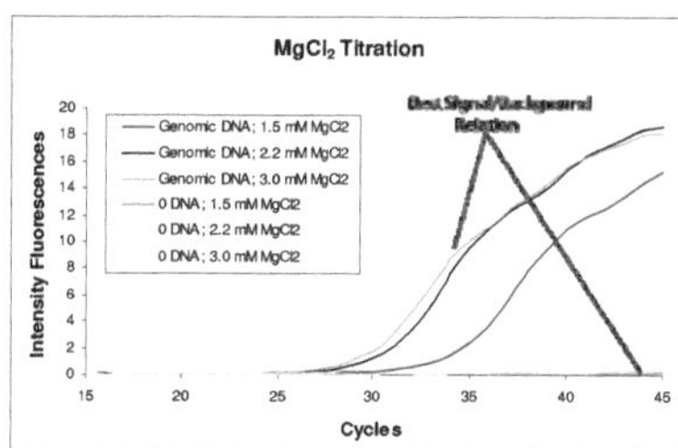


**Fig. 11: $MgCl_2$ titration experiment using *S. pombe* genomic DNA.**

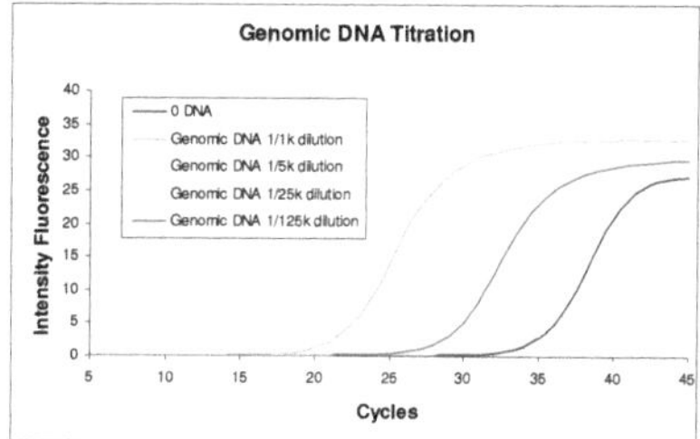


**Fig. 12: Dilution experiment using S. pombe genomic DNA**

In order to confirm that the found melting temperature corresponds to the amplified product, a fraction of DNA was loaded on a 2% agarose gel (results not shown).

# Chapter 4

# Results

## 4.1 Cell Lysate Purification / CsCl Gradient

After CsCl gradient centrifugation 12 fractions of cell lysate were obtained. Fractions 4 to 10 were pooled by the criteria mentioned in the materials chapter. Those 7 fractions represent almost 75 % of the total crosslinked DNA. Therefore through CsCl gradient centrifugation one looses 25 % of crosslinked material but this is outweigh by the benefit of lysate purification.

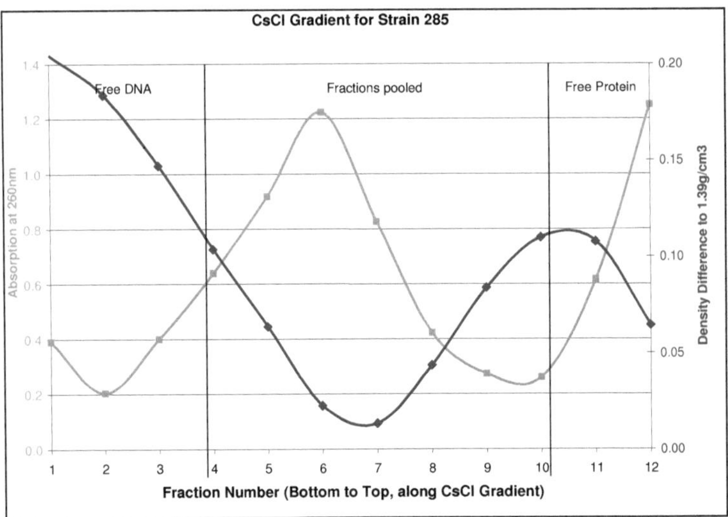


**Fig. 13: Lysate purification via CsCl gradient centrifugation.**
To the obtained lysate CsCl was added and ultra-centrifuged for 72 hours. 12 fractions were separated, from bottom to top labeled 1 – 12. The Absorbance at 260 nm (mainly DNA, red line), 280 nm (mainly protein) and their ratio was taken. Together with density measurements the optimum fractions to pool were determined to be fraction 4 to 10. Since DNA-protein crosslinked fractions have a density of app. 1.39 g/cm3, the plot of variation from that value allows selecting a reasonable cutoff.

| Bottom <-- | | | | | | Gradient | | | | | --> Top | |
|---|---|---|---|---|---|---|---|---|---|---|---|---|
| Fraction | 1 | 2 | 3 | 4 | 5 | 6 | 7 | 8 | 9 | 10 | 11 | 12 | Total |
| DNA [ug] | 7.8 | 4.1 | 8.0 | 12.8 | 18.4 | 24.5 | 16.5 | 8.5 | 5.5 | 5.2 | 12.3 | 25.0 | 148.5 |
| Percentage of DNA pooled over all: 54.3 | | | | | | | | | | | | | |
| Percentage of cross-linked DNA pooled: 72.5 | | | | | | | | | | | | | |

**Table 3: Purification of Cell lysate by excluding fractions with high concentration of non-crosslinked proteins and free DNA.**
To obtain the best yield but purify the sample enough, fractions 4 to 10 were pooled. The top fractions 11 and 12 mainly contain free protein (which also misleads to high absorption readings at 260 nm) and the bottom fraction uncrosslinked DNA. Therefore it can be assumed that the total recovery rate of crosslinked DNA is over 70%.

## 4.2 ORC binds to Origins of Replication

Origins of replication are the places at which DNA replication starts. The Origin Recognition Complex (ORC) is the first of several proteins assembling at origins of replication during $G_1$ phase. In order to proof that ORC is present at origins of replication a ChIP assay was performed as described under Methods. The mutant used in this experiment (see table 1), JLP285, contained the *GFP* gene fused to the *orp2* gene. The expression of the tagged *orp2* is under the control of its native regulatory elements so as to achieve wildtype levels of expression. The phenotype of JLP285 does not differ from the wildtype phenotype, suggesting that Orp2-GFP can functionally replace the wildtype protein. Since the whole origin recognition complex is maintained throughout the cell cycle in *S. pombe,* GFP fused to the second subunit of the ORC complex allows precipitating the entire complex. The successful co-immunoprecipitation of ORC with origin DNA (see figure 14) is the basis for further experiments as it shows that the ChIP assay used in the lab can reproduce previously published results. It also solidifies the current model of initiation of replication in *S. pombe*.

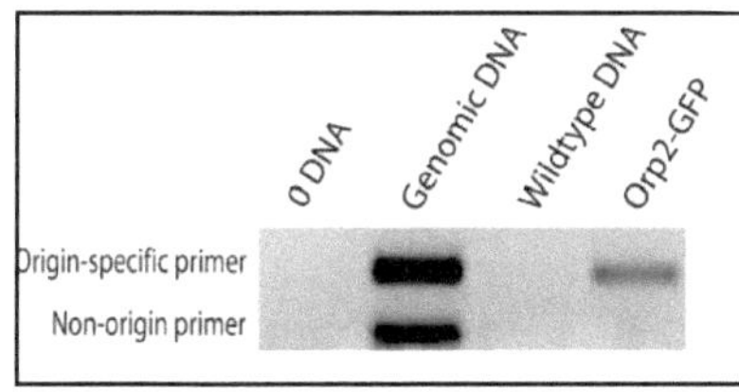


**Fig. 14: ORC binds to Origin of Replication in vivo.**
A Chromatin Immunoprecipitation assay was performed as described in Material & Methods. The PCR products were analyzed via electrophoresis on a 2% agarose gel.

As controls for the PCR a negative control, lacking any DNA and a positive control, containing genomic DNA were included to ensure that immunoprecipitation of origin

DNA is not a phenomenon of the fused GFP itself. As an additional negative control a ChIP assay with a wildtype strain was included. No signal was obtained using wildtype template for neither the origin-specific nor the non-origin primers. In contrast the *Orp2-GFP* mutant template yielded a clear signal when amplified with the origin-specific primer pair. The absence of a signal employing the non-origin primer verifies that the signal is indeed related to ORC-GFP at origin DNA and not an event of mispriming.

Several replication origins have been identified in fission yeast by their ability to act as autonomously replicating sequences (ARS) (Dubey et al., 1994; Okuno et al., 1999). These are composed of adenine- and thymidine- enriched regions of DNA clustered within a 1-2 Kb region of DNA. In this research a specific region of the ARS3001 origin, labeled as region 9 was used. It is utilized as replication origin, binds components of both the ORC (see results above) and MCM complexes (personal communication J. Leatherwood), and lies within the ribosomal repeats. This ensures higher copy numbers and a greater yield of immunoprecipitated DNA than it would be in ARS regions that lie outside the ribosomal repeats, such as the commonly used ARS3002 and ARS2004.

For the ChIP analysis the specific ARS3001-reg9 origin primers, (9F/9R) and non-specific control primers ARS3001-reg17 (17F/17R) were used. The origin specific primer recognizes the ~90bp origin segment across the ARS3001 locus, whereas the control primer amplifies a 70 bp region located 8-10 Kb away at the 5' respectively 3' end of ARS3001-reg9 (since ARS and non-ARS regions alternate within the ribosomal repeats).

Chromosome III in *S .pombe* contains app. 100–150 copies of the 10.9 Kb rDNA gene cluster (17 S, 5.8 S and 25 S rRNA). Previous studies have shown that each non-transcribed spacer region contains one copy of ARS3001 (Figure 15), a comparatively small *S. pombe* replication origin of 573 bp that contains four genetically required DNA regions. Previous studies have also shown that the *S. pombe* Orp4 protein is solely responsible for binding the *S. pombe* ORC to specific AT-rich sites within *S. pombe* replication origins (see also figure 3). Orp4 binds to the Δ3 and Δ6 regions, but not to either the Δ2 or Δ9 regions. Moreover, the affinity of Orp4 for Δ3 is about six times greater than for Δ6 (Kim and Huberman, 1998). Therefore, a primer pair was used which is designed to amplify the 299 bp sequence of region Δ 2 and Δ 3 (which are adjacent).

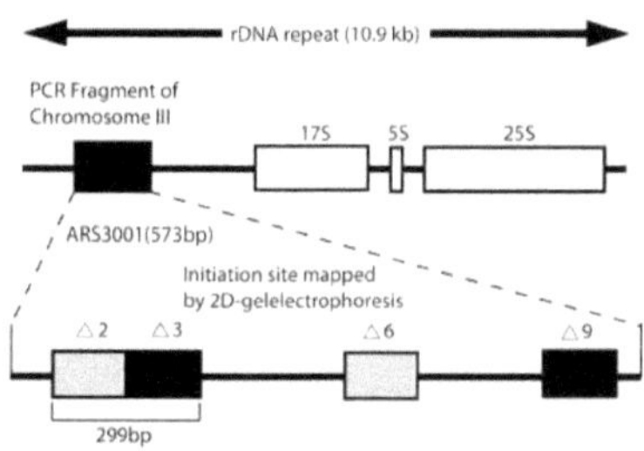


**Fig. 15: Initiation site of ARS3001, as used in this study.**
Orp4 binds to two sites, Δ3 and Δ6, in ARS3001. ARS3001 is located in the non-transcribed spacer of the rDNA repeats and contains two strongly required DNA regions (Δ3, Δ9) and two moderately required DNA regions (Δ2, Δ6) (Kim and Huberman, 1998). Using an origin of replication within the ribosomal repeats with high affinity of ORC ensures the highest possible copy number of ORC-DNA complexes to investigate.

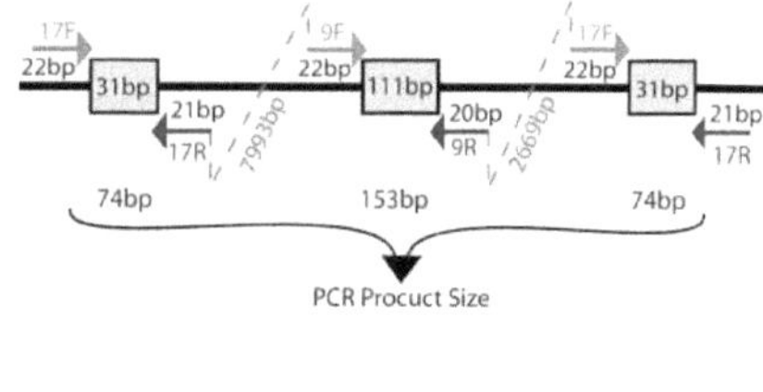


**Fig. 16: Schematic view of primers and their products amplified at ARS3001.**
The showed region is a magnification of the ARS3001 region as shown in figure 15. The origin specific 9F/R primer pair amplifies the Δ3 region of ARS3001 and yields a PCR product of 153 bp length. App. 8 Kb downstream and ~2.7 Kb upstream to the Δ2/Δ3 region, regions that are known not to act as origin of replication are amplified by the 17F/R control primer pair. The obtained PCR product is a 74 bp fragment.

## 4.3 Cdc2-GFP co-immunoprecipitates with origin DNA

Although ORC binds constitutively to chromatin throughout the fission yeast cell cycle, the prereplicative complex only forms on origins in $G_1$. So presumably there is a factor, which inhibits reformation of the pre-RC once origins are fired. This factor might stably bind ORC during $G_2$ and early M phase to prevent reinitiation of S phase. Postulating that this factor is the Cdc13/Cdc2 complex itself a GFP-tagged *cdc2* mutant and chromatin immunoprecipitation (ChIP) analysis were used to identify a possible binding of Cdc2 to ORC. The GFP is attached at the C-terminus Cdc2 and since the cells do not show any mutant phenotype the assumption can be made, that the GFP tag does not affect Cdc2 function in vivo.

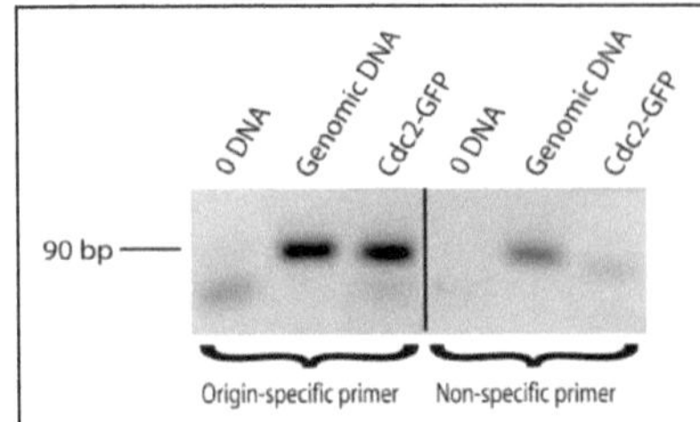


**Figure 17: Cdc2-GFP co-immunoprecipitates with origin DNA.**
A ChIP assay was performed as described in Methods. Fragmented chromosomal DNA was purified and the target sequence amplified by real-time PCR. The product was analyzed via electrophoresis on a 2% agarose gel.

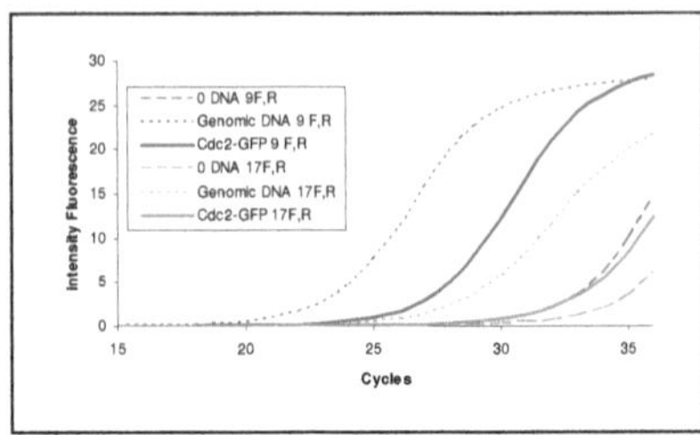


**Figure 18: Quantification of Cdc2-GFP.**
The Cdc2-GFP signal (solid line) obtained by origin-specific priming with a half-maximum value of 27.07 cycles is clearly above background (dashed line, >32 cycles).

As controls for the real-time PCR a negative control, lacking any DNA and a positive control, containing genomic DNA were included. Origin-specific primers and non-origin primers were employed as mentioned above. The *cdc2-GFP* mutant template yielded a clear signal when amplified with the origin-specific primer pair. The weak unspecific signal employing the non-origin primer is not related to any IP product as the melting curve analysis verified. The signal obtained with the origin specific primers is related to Cdc2-GFP at origin DNA and not an event of mispriming, since a similar signal is lacking in the Cdc2-GFP (non-specific primer) lane as shown in figure 17.

# Chapter 5

# Discussion

## 5.1 Development of a Working ChIP Assay for Routinely Lab Use

The successful development of a working ChIP assay is a win for every researcher investigating DNA / protein or protein / protein interactions. It broadens the spectrum of experiments that can be conducted and gives the investigator a powerful tool to observe those interactions in vivo.

Starting with existing protocols, those protocols could be further developed for higher contrasting results and more sensitive product detection. Three major modifications, respectively additions, were successfully implemented. Firstly, the optimization of lysate sonication led to smaller DNA fractions without loosing product due to overheating and degradation. This in turn reduced precipitation of non-origin regions with the GFP-antibody. The result is an enhanced signal to background ratio and a more stringent detection of GFP-tagged proteins. Secondly the implementation of an additional cleanup step in form of the CsCl density centrifugation further enhanced the specificity of the ChIP assay. Remaining uncrosslinked DNA and free protein could be reduced to a minimum without a significant loss of the desired crosslinked DNA - protein, respectively protein - protein material. As a third major modification the product detection via real time PCR was implemented. Contrasting regular PCR, real time PCR using LaRoche's LightCycler is not only faster but also supplies the investigator with more information about the product(s) amplified in the reaction. By simultaneously obtaining a melting curve of the products one can instantly distinguish between the desired product and unspecific background product.

## 5.2 Verifying Previous Findings of ORC – Origin Interaction

Obtaining a positive result for the ORC - origin interaction, employing the ChIP assay is a positive control for the ChIP procedure itself. The strong signal for the ORC-GFP mutant in contrast to the absent signal of the wildtype strain shows that the chosen conditions and procedures work excellent for ChIP. The ChIP assay can therefore be implemented into future experiments in the lab.

Furthermore, the findings of other groups (reviewed in Bell and Dutta 2002) that in fission yeast ORC does bind to origins of replication in vivo were verified. The results presented in this projectwork are also congruent with experiments done in budding yeast (*S. cerevisiae*), flies (*Drosophila melanogaster, Sciara coprophila*) and mammals (Abdurashidova 2003, Bell 2002, Keller 2002, Kong and DePamphilis 2002, Ladenburger 2002). In *S. cerevisiae*, an essential role for ORC in DNA replication is the recruitment of additional replication control proteins to yeast origins of replication, including Cdc6p, the MCM proteins, and Cdc45p (Aparicio et al. 1997; Tanaka et al. 1997). Because sequence analogs for most of these replication control proteins have been identified in *S. pombe*, it is likely that ORC will direct replication by recruiting these proteins to *S. pombe* origins of DNA replication. The experiments described in this projectwork show that ORC specifically binds to the ARS3001 origin, but studies of ORC localization suggest mechanisms must also exist to prevent ORC from binding to nonamplifying loci. These mechanisms could include cofactors that possess chromatin remodeling activity or that physically interact with ORC to restrict binding to the amplified regions.

The in vivo methods described here for studying ORC-DNA interactions in *S. pombe* cells can be adapted to identify additional ORC DNA-binding sites and potential replication origins. Since fission yeast origins have no consensus sequence like *S. cerevisiae* does, identification of these sequences will be important for understanding how ORC regulates replication and how origin usage is regulated. Using these sequences in combination with in vitro replication systems (Crevel and Cotterill 1991, Walter et al. 1998; Chesnokov et al. 1999) and the future purification of ORC will allow for the design of more precise experiments to understand how DNA replication is regulated.

## 5.3 Showing Cdc2 – ORC interaction in vivo

Genetic studies of fission yeast have established that the cyclin-dependent kinase Cdc2 is required for both positive and negative regulation of DNA replication. These studies have shown that elimination of Cdc2 activity in G1 phase prevents the onset of S phase, whereas elimination of Cdc2 kinase in G2 phase causes the reinitiation of S phase without an intervening M phase (Broek et al., 1991; Hayles et al., 1994; Moreno and Nurse, 1994; Nurse and Bissett, 1981). The goal of this investigation has been to discover if ORC and Cdc2 kinase are localized at DNA replication origins as predicted by current models and if ORC binds Cdc2 in vivo.

Since ORC is constitutively bound to replication origins throughout the fission yeast cell cycle and Orp2 interacts with Cdc2 by two hybrid analysis, the data shown in this work suggest Orp2 provides a docking site for Cdc13/Cdc2 at replication origins (Leatherwood et al., 1996; Lygerou and Nurse, 1999). Furthermore, mutational analysis of Cdc2-dependent phosphorylation sites has suggested that phosphorylation of Orp2 by Cdc2 contributes to suppression of pre-RC formation during a normal cell cycle (Vas et al., 2001). Importantly, cells that express a mutant Orp2 protein which substantially abrogates binding of Cdc13 to replication origins are acutely prone to diploidization, but this is not observed in cells in which all of the known Cdk phosphorylation sites in Orp2 have been abolished. From the obtained results in this work it can be proposed that ORC bound Cdc13/Cdc2 kinase acts locally at the replication origin to prevent pre-RC formation by phosphorylation of additional proteins such as Orc6, Cdc18, and possibly other components of the pre-RC complex. To formally verify this prediction, it would be necessary to simultaneously mutate all the phosphorylation sites on each of these proteins, as has been attempted previously in budding yeast (Nguyen et al., 2001).

# Chapter 6

# Summary and Outlook

Having ORC physically bound to origins of replication suggest that this is the premier regulatory step in the assembly of pre-RCs and thereby the commitment to DNA replication. Overall the analysis of co-immunoprecipitation of ORC with DNA sequences acting as origins of replication provides strong in vivo evidence supporting the hypothesis that ORC acts early in S phase in general, and functions as the initiator protein in fission yeast at origins of replication in particular. From another point of view this also means that ORC marks the sites of replication in the chromosome. The involvement of six proteins in this complex, however, suggests that ORC also performs other functions at the origin. By analogy to known initiator proteins, such a function could be directing the assembly of other DNA replication proteins to the origin. The idea of ORC acting as a "landing pad" for proteins of the pre-RC was suggested previously (Prasanth et al., 2002) and explains nicely the early recruitment of ORC to DNA. The complex nature of eukaryotic chromosomal replication suggests a number of other possible functions for ORC, including the localization of origins of replication within the nucleus, assisting the chromatin assembly process, or maintaining a chromatin-free region at the origin of replication. ORC also presents a target for cell cycle regulators that control the G1 to S phase transition and the replication of the genome once per cell cycle. For fission yeast this was first shown in vitro in a two hybrid interaction (Leatherwood et al., 1996). The presented results confirm these findings in vivo, using a more direct approach of investigation. Further genetic analysis of the genes encoding the subunits of ORC as well as continued biochemical analysis of ORC should provide insights into both the functions of ORC during yeast DNA replication and the biochemical mechanisms through which they are exerted.

Recent evidence suggests metazoans have evolved additional mechanisms to prevent relicensing of origins to those found in the yeasts. Whereas the ORC complex remains bound to replication origins throughout the fission yeast cell cycle, in human cells the Orc1 protein undergoes phosphorylation-induced degradation after initiation of DNA replication (Lygerou and Nurse, 1999; Méndez et al., 2002). This may be an additional event catalyzed by a Cyclin/Cdk complex bound to the remaining ORC subunits at replication origins in human cells. More particularly, human and frog cells contain a protein called geminin that inhibits the activity of the Cdt1 loading factor and that, like cyclin B, is destroyed at anaphase (McGarry and Kirschner, 1998; Wohlschlegel et al., 2000; Tada et al., 2001). Geminin may operate to prevent ectopic formation of prereplicative complexes in G2. However, immunodepletion of geminin from cycling *Xenopus* extracts does not cause the repetitive initiation of S phase, suggesting that other mechanisms are involved. Fission yeast, which also replicates its chromosomes only once per cell cycle, does not contain an obvious geminin homolog. This argues that stable association of the mitotic Cyclin B/Cdc2 kinase to replication origins may be the primary mechanism by which the order of S phase and M phase is maintained during the mitotic cell cycle of eukaryotes.

For future experiments it would be challenging to address the question of how the ORC-DNA and Cdc2-ORC interaction changes over the course of a complete cell cycle. When exactly do these interactions happen and what gene products are involved in its regulation? One very elegant way to answer those questions would be a time course Microarray experiment, using the ChIP assay for each single time point. Since in a Microarray experiment all 4929 *S. pombe* genes are probed at once, this "ChIP on a chip" could also reveal yet undiscovered origins of replication in one single experiment. Furthermore, genes that show high transcriptional activity in coincidence with ORC-DNA and Cdc2-ORC interaction could be identified and further investigated for their role in DNA replication.

## Acknowledgements

First of all I would like to thank Dr. Janet Leatherwood who not only supervised my experiments in the best way imaginable, but who also always had the time and patience to guide me through the process of becoming a scientific researcher. She also enabled me to get financial support for this work through the Stony Brook University Research Foundation.

I am grateful for Nicole Averbeck's support, encouragement, and critical reading of the manuscript. I also would like thank her for the many pieces of good advice during the daily lab routine which were always helpful and settled a lot of problems right away.

I am thankful for Sum Yip's expert technical assistance, Dr. Janet Hearing's help with the gradient centrifugation, and our laboratory group's valuable discussions about my work.

A special thanks goes to Doreen for her great patience, loving support, and trust throughout the course of this project.

## Figures

# Tables

# References

Bell, S. P., and Dutta, A. (2002). DNA replication in eukaryotic cells. Annu Rev Biochem *71*, 333-374.

Booher, R. N., Alfa, C. E., Hyams, J. S., and Beach, D. H. (1989). The fission yeast cdc2/cdc13/suc1 protein kinase: regulation of catalytic activity and nuclear localization. Cell *58*, 485-497.

Broek, D., Bartlett, R., Crawford, K., and Nurse, P. (1991). Involvement of p34cdc2 in establishing the dependency of S phase on mitosis. Nature *349*, 388-393.

Correa-Bordes, J., and Nurse, P. (1995). p25rum1 orders S phase and mitosis by acting as an inhibitor of the p34cdc2 mitotic kinase. Cell *83*, 1001-1009.

Diffley, J. F. (2001). DNA replication: building the perfect switch. Curr Biol *11*, R367-370.

Dubey, D. D., Zhu, J., Carlson, D. L., Sharma, K., and Huberman, J. A. (1994). Three ARS elements contribute to the ura4 replication origin region in the fission yeast, Schizosaccharomyces pombe. Embo J *13*, 3638-3647.

Dutta, A., and Bell, S. P. (1997). Initiation of DNA replication in eukaryotic cells. Annu Rev Cell Dev Biol *13*, 293-332.

Hayles, J., Fisher, D., Woollard, A., and Nurse, P. (1994). Temporal order of S phase and mitosis in fission yeast is determined by the state of the p34cdc2-mitotic B cyclin complex. Cell *78*, 813-822.

Hendrickson, M., Madine, M., Dalton, S., and Gautier, J. (1996). Phosphorylation of MCM4 by cdc2 protein kinase inhibits the activity of the minichromosome maintenance complex. Proc Natl Acad Sci U S A *93*, 12223-12228.

Hochstenbach, F., Klis, F. M., van den Ende, H., van Donselaar, E., Peters, P. J., and Klausner, R. D. (1998). Identification of a putative alpha-glucan synthase essential for cell wall construction and morphogenesis in fission yeast. Proc Natl Acad Sci U S A *95*, 9161-9166.

Huxley, T. H. (1869). An Introduction to the Classification of Animals.

Kearsey, S. E., Montgomery, S., Labib, K., and Lindner, K. (2000). Chromatin binding of the fission yeast replication factor mcm4 occurs during anaphase and requires ORC and cdc18. Embo J *19*, 1681-1690.

Kelly, T. J., and Brown, G. W. (2000). Regulation of chromosome replication. Annu Rev Biochem *69*, 829-880.

Kim, S. M., and Huberman, J. A. (1998). Multiple orientation-dependent, synergistically interacting, similar domains in the ribosomal DNA replication origin of the fission yeast, Schizosaccharomyces pombe. Mol Cell Biol *18*, 7294-7303.

Lang, B. F., O'Kelly, C., Nerad, T., Gray, M. W., and Burger, G. (2002). The closest unicellular relatives of animals. Curr Biol *12*, 1773-1778.

Leatherwood, J., Lopez-Girona, A., and Russell, P. (1996). Interaction of Cdc2 and Cdc18 with a fission yeast ORC2-like protein. Nature *379*, 360-363.

Lee, K. M., Saiz, J. E., Barton, W. A., and Fisher, R. P. (1999). Cdc2 activation in fission yeast depends on Mcs6 and Csk1, two partially redundant Cdk-activating kinases (CAKs). Curr Biol *9*, 441-444.

Lindner, P. (1893). Wochenschrift fuer Brauerei *10*, 1298-1300.

Lopez-Girona, A., Mondesert, O., Leatherwood, J., and Russell, P. (1998). Negative regulation of Cdc18 DNA replication protein by Cdc2. Mol Biol Cell *9*, 63-73.

Moreno, S., Hayles, J., and Nurse, P. (1989). Regulation of p34cdc2 protein kinase during mitosis. Cell *58*, 361-372.

Moreno, S., and Nurse, P. (1994). Regulation of progression through the G1 phase of the cell cycle by the rum1+ gene. Nature *367*, 236-242.

Nguyen, V. Q., Co, C., and Li, J. J. (2001). Cyclin-dependent kinases prevent DNA re-replication through multiple mechanisms. Nature *411*, 1068-1073.

Nurse, P. (1990). Universal control mechanism regulating onset of M-phase. Nature *344*, 503-508.

Nurse, P., and Bissett, Y. (1981). Gene required in G1 for commitment to cell cycle and in G2 for control of mitosis in fission yeast. Nature *292*, 558-560.

Okuno, Y., Satoh, H., Sekiguchi, M., and Masukata, H. (1999). Clustered adenine/thymine stretches are essential for function of a fission yeast replication origin. Mol Cell Biol *19*, 6699-6709.

Patterson, D. J. (1999). The Diversity of Eukaryotes. Am Nat *154*, S96-S124.

Prasanth, S. G., Prasanth, K. V., and Stillman, B. (2002). Orc6 involved in DNA replication, chromosome segregation, and cytokinesis. Science *297*, 1026-1031.

Sambrook, J., Fritsch, E.F., Maniatis, T (1989). Molecular Cloning - A Laboratory Manual, Second edn, Cold Spring Harbor Laboratory Press).

Strahl-Bolsinger, S., Hecht, A., Luo, K., and Grunstein, M. (1997). SIR2 and SIR4 interactions differ in core and extended telomeric heterochromatin in yeast. Genes Dev *11*, 83-93.

Thein, S. L., Lynch, J. R., Weatherall, D. J., and Wallace, R. B. (1986). Direct detection of haemoglobin E with synthetic oligonucleotides. Lancet *1*, 93.

Tye, B. K. (1999). MCM proteins in DNA replication. Annu Rev Biochem *68*, 649-686.

Vas, A., Mok, W., and Leatherwood, J. (2001). Control of DNA rereplication via Cdc2 phosphorylation sites in the origin recognition complex. Mol Cell Biol *21*, 5767-5777.

Wood, V., Gwilliam, R., Rajandream, M. A., Lyne, M., Lyne, R., Stewart, A., Sgouros, J., Peat, N., Hayles, J., Baker, S., *et al.* (2002). The genome sequence of Schizosaccharomyces pombe. Nature *415*, 871-880.